# ABOUT
# SPACE

By Jana Carson

**TREASURE BAY**
Family Engagement in Reading

# Parent's Introduction

**Whether your child is a beginning reader, a reluctant reader, or an eager reader, this book offers a fun and easy way to encourage and help your child in reading.**

Developed with reading education specialists, **We Both Read** books invite you and your child to take turns reading aloud. You read the left-hand pages of the book, and your child reads the right-hand pages—which have been written at one of six early reading levels. The result is a wonderful new reading experience and faster reading development!

You may find it helpful to read the entire book aloud yourself the first time, then invite your child to participate the second time. As you read, try to make the story come alive by reading with expression. This will help to model good fluency. It will also be helpful to stop at various points to discuss what you are reading. This will help increase your child's understanding of what is being read.

In some books, a few challenging words are introduced in the parent's text, distinguished with **bold** lettering. Pointing out and discussing these words can help to build your child's reading vocabulary. If your child is a beginning reader, it may be helpful to run a finger under the text as each of you reads. Please also notice that a "talking parent" ☺ icon precedes the parent's text, and a "talking child" ☺ icon precedes the child's text.

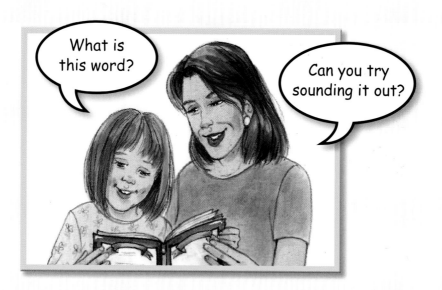

If your child struggles with a word, you can encourage "sounding it out," but keep in mind that not all words can be sounded out. Your child might pick up clues about a word from the picture, other words in the sentence, or any rhyming patterns. If your child struggles with a word for more than five seconds, it is usually best to simply say the word.

Most of all, remember to praise your child's efforts and keep the reading fun. At the end of the book, there is a glossary of words, as well as some questions you can discuss. Rereading this book multiple times may also be helpful for your child.

Try to keep the tips above in mind as you read together, but don't worry about doing everything right. Simply sharing the enjoyment of reading together will increase your child's reading skills and help to start your child off on a lifetime of reading enjoyment!

About Space
Fourth Edition

A We Both Read Book
Level 1–2
Guided Reading: Level H

*With special thanks to Maria Alvarellos at the Lawrence Hall of Science, University of California at Berkeley, for her review and recommendations on the material in this book.*

Images courtesy of NASA, NSSDC, NASA/JPL-Caltech, and NASA/JSC.
Image of Keck telescopes on page 36 courtesy of W. M. Keck Observatory.

We Both Read® is a registered trademark of Treasure Bay, Inc.

Published by
Treasure Bay, Inc.
P.O. Box 119
Novato, CA 94948 USA

Printed in Malaysia

Library of Congress Control Number: 2013942595

ISBN: 978-1-60115-022-6

Visit us online at:
WeBothRead.com

PR-11-19

# TABLE OF CONTENTS

Carina Nebula

Let's take a journey into space, where we will see wondrous sights and make incredible discoveries.

What is space? Space is the **universe**. The **universe** contains everything—including all the stars and planets and all the empty expanses in between them.

Have you ever looked up at the sky? It seems so big and there are so many stars! Yet what we see in our sky is only a very tiny part of the **universe**.

A galaxy is a large system of stars held together in a group by gravity. We are in a galaxy called the **Milky Way**, and the stars that we are able to see in our night sky are part of our galaxy. We can see many more of the stars in our galaxy through the use of a telescope.

*Milky Way galaxy*

Long ago, people could only look at the stars with their eyes. Most of the stars of the **Milky Way** just looked like a white streak in the sky.

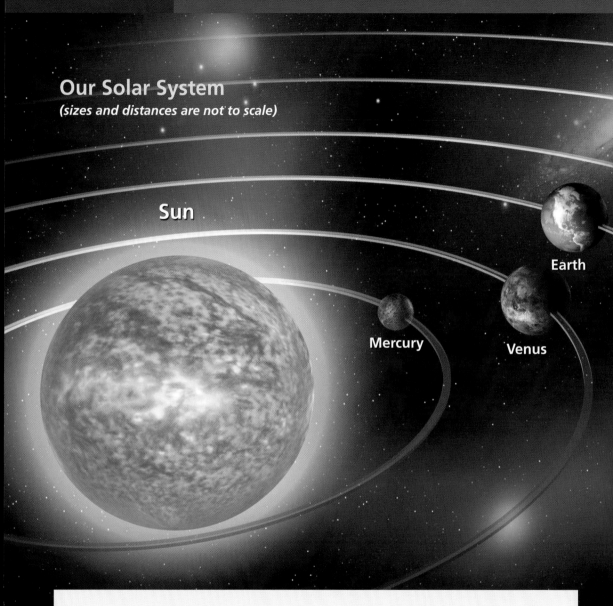

Our Solar System
*(sizes and distances are not to scale)*

Sun

Earth

Mercury

Venus

Within a galaxy there may be many **solar systems**. A **solar system** is made up of a sun and everything that moves around it. Our **solar system** exists within the Milky Way galaxy. It includes all the planets and their moons as well as the comets, asteroids, and space objects that orbit, or move in circles, around our sun.

Uranus

Neptune

Jupiter

Saturn

Mars

A star with planets around it is called a sun. There is a sun at the center of every **solar system**. Without our sun, there could be no life in our **solar system**.

The planet we live on is called Earth. As far as we know, it is the only planet in our solar system that has life on it. Other planets in our solar system are **Mercury**, Venus, Mars, Jupiter, Saturn, Uranus, and Neptune.

**Mercury** is the closest planet to our sun. The temperature on the planet's **surface** is hot enough to melt a tin pan.

*Closer view of Mercury showing craters and the great Caloris basin (large tan area)*

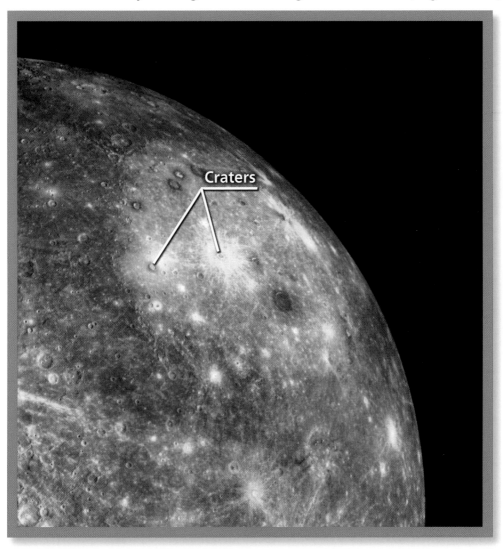

Craters

Mercury is much smaller than our planet. There is no water or air on the **surface** of Mercury, and it is very hot.

The surface of Mercury looks a lot like our moon. There are many craters.

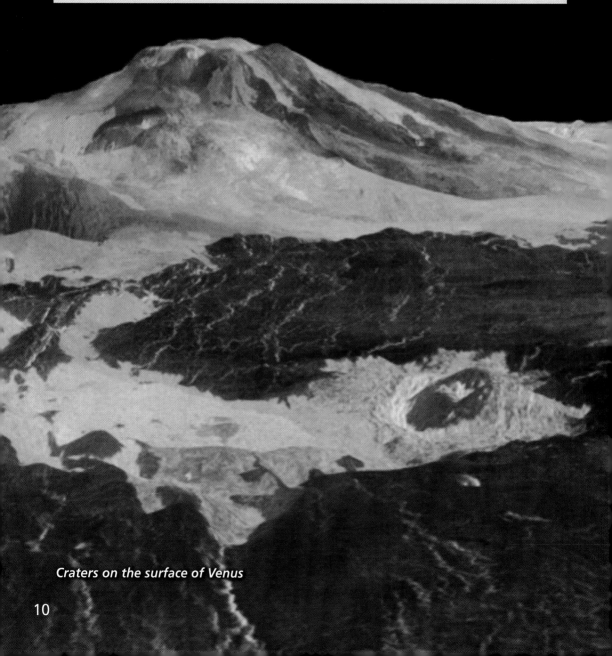

Venus and Earth are similar in size, and they both have mountains and valleys and plains. However, there is no water on **Venus**. Scientists believe there may have been water there billions of years ago, but it has all now boiled off into steam or **clouds**. Venus is also known as the "**evening star**."

*Craters on the surface of Venus*

 **Venus** has been called the **evening star** because it shines very brightly at just about sunset. Venus is covered with thick **clouds**. There are always very big storms in these clouds.

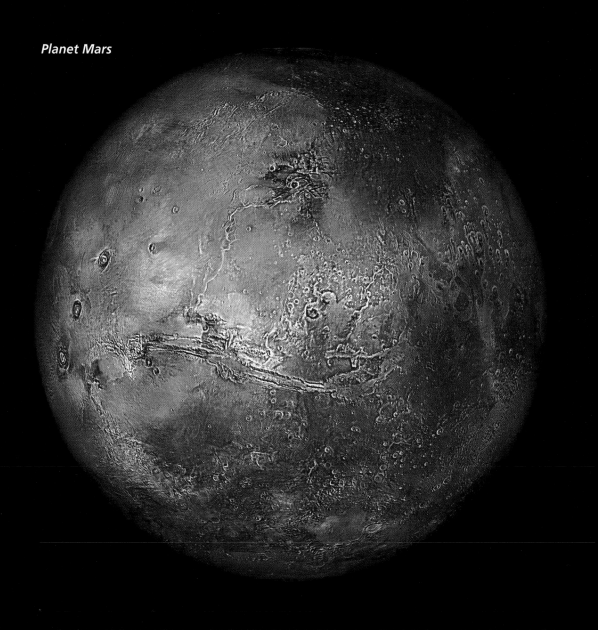

 **Mars** is often called the Red Planet. **Space rovers** have been sent to **Mars** by the United States and other countries. The **space rovers** collect information that is examined by **scientists**. The **space rovers** found a lot of metal called iron in the soil on **Mars**. The iron is what gives the planet its reddish color.

*NASA's space rover, Curiosity, on the surface of Mars*

The **space rovers** have found many other things. Some things have led **scientists** to think that long ago there could have been some form of life on **Mars**.

Huge aurora on Jupiter photographed
by the Hubble telescope

Great Red Spot

⊝⃝  **Jupiter** is the largest planet in our solar system. There
are terrific lightning bolts and huge gas storms in **Jupiter's**
atmosphere.

A large area of swirling gas called the Great Red Spot is
believed to be a hurricane-like storm.

Size Comparison

Jupiter          Earth

Jupiter is a huge planet. It is so big, that all the other planets in our solar system could fit inside it.

Jupiter has more than sixty moons. One is almost as big as the planet Mars!

**Saturn** is the second-largest planet in our solar system. This planet spins rapidly on its axis, causing the top and bottom of the planet to flatten out.

Many moons **orbit** Saturn. The largest is called Titan. Some scientists think there may be a salty ocean beneath the surface of Titan.

*Saturn's rings*

 There are also rings that **orbit Saturn**. These rings are made up of ice and small rocks. Some chunks of ice are as big as a house. Some are as small as a speck of dust.

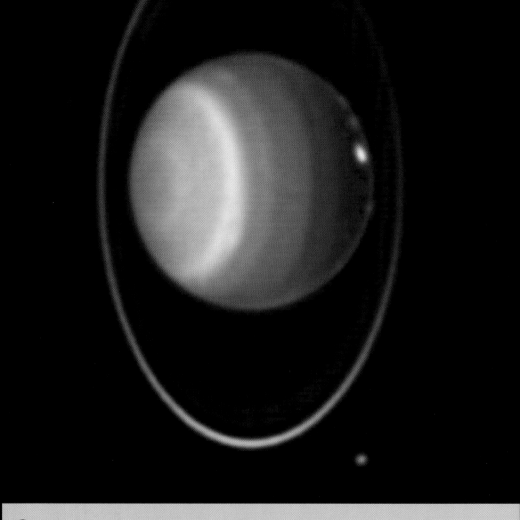

Uranus and **Neptune** have similar atmospheres composed primarily of hydrogen and helium gases. However, Uranus is unique because of how it is tilted on its axis. It lies almost on its side in relation to the sun. When the sun rises at its north pole, it stays up for forty-two Earth years before it sets!

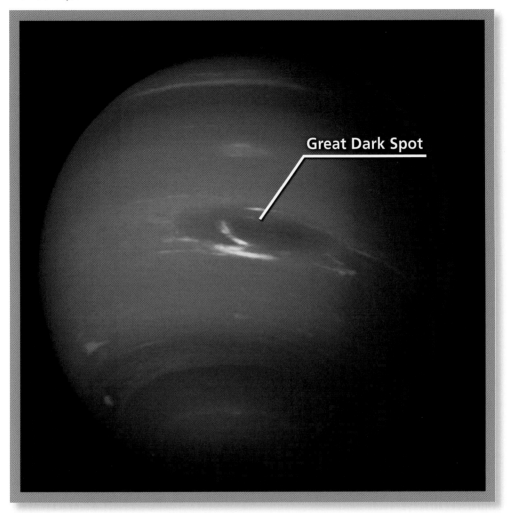

Great Dark Spot

Both of these planets have rings around them. The rings around **Neptune** are very hard to see.

Neptune has huge storms on the surface. These storms can be seen on the planet as dark spots. They are called Great Dark Spots, and they can be as wide across as our entire planet.

*Dwarf planet Pluto*

Pluto was once considered a planet. However, the discoveries of other natural objects in our solar system similar in size to **Pluto** have caused **astronomers** to reconsider what the term *planet* should mean.

After much debate, many **astronomers** agreed that **Pluto** and some other large, natural objects in our solar system belong in a new category named "**dwarf** planets."

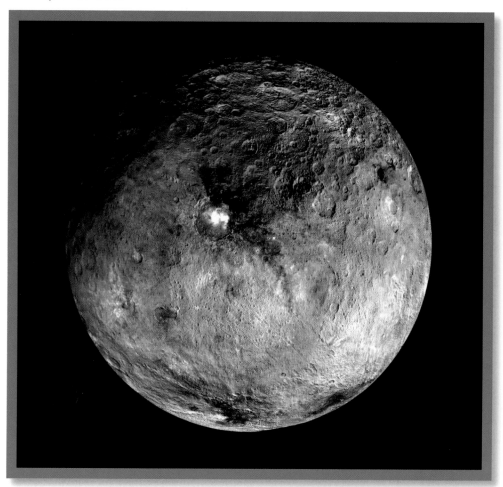

**Astronomers** found a **dwarf** planet that is about the same size as **Pluto**. It is called Eris, and it is much farther from the sun than Pluto.

Astronomers decided that Ceres (SEER-eez) should be called a dwarf planet. Ceres is much smaller than Earth's moon, and it orbits the sun between Mars and Jupiter.

## Size Comparison

**Earth**            **Moon**

**Earth** is our home planet. It's the third planet from the sun and is the only planet in our solar system that has flowing water on its surface.

About seventy percent of **Earth's** surface is covered with water. Continents with mountains, plains, forests, and **deserts** cover the remaining thirty percent.

**Earth** has one moon. Our moon is like a very dry **desert**. From here on Earth we can only see one side of our moon. The side that we see is called the near side. The side we never see is called the far side.

Our moon has had visitors! In 1969, the *Apollo 11* spaceship carried astronauts Neil Armstrong and Edwin "Buzz" Aldrin to the moon to explore its surface. Neil Armstrong was the first person to walk on the moon.

He left his footprints. There is no air on the moon to blow them away. His footprints are still on the moon.

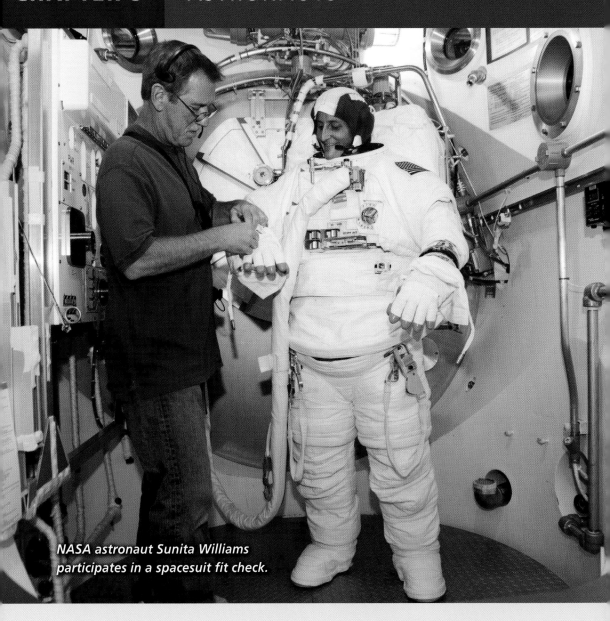

*NASA astronaut Sunita Williams participates in a spacesuit fit check.*

**Astronauts** go through years of specialized training. They must have strong skills in science, math, and technology.

The **astronauts** that go into space must learn how to function in weightless environments and even learn how to do a spacewalk.

Sometimes **astronauts** train underwater. The water helps them know what it feels like to float in space.

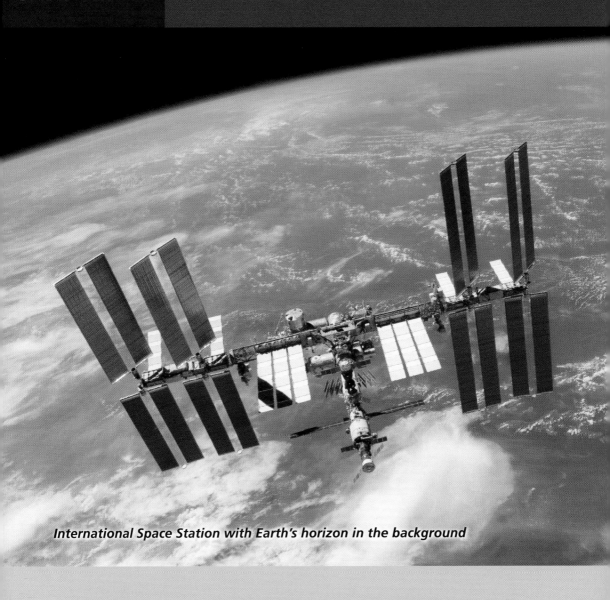

*International Space Station with Earth's horizon in the background*

How would you like to live in space? Some astronauts do. There are teams of astronauts that take turns living and working on a space station. A space station is an enormous satellite that orbits Earth.

The biggest space station so far is the **International Space Station**.

*Astronauts doing construction and maintenance on the International Space Station*

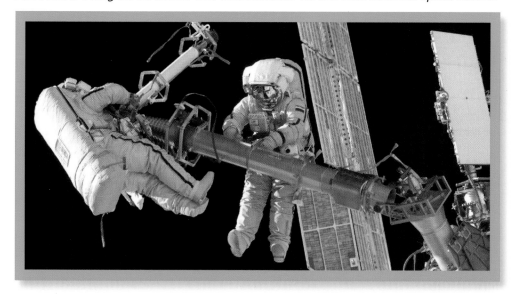

The **International Space Station** is made up of many different parts. Astronauts put these parts together in space.

*Astronauts, fruit, vegetables, and a plant floating in the International Space Station*

Many types of science **experiments** done on the space station cannot be done on Earth. There is no **gravity** on the space station. Everything that is not attached to the inside of the space station floats freely in the air.

Some **experiments** explore the effects of no **gravity**. For example, how do plants grow when there is no gravity? Or what is the effect on humans and animals when they live without gravity for weeks or months?

Mealtime aboard the International Space Station

Living on a space station has its own special challenges. Astronauts do get some fresh fruits and vegetables, but some of the food the astronauts eat has been freeze-dried—a special process used to remove all of the water from the food. Before astronauts eat their freeze-dried food, they put the water back in it.

Just like many kids, astronauts drink from boxes or pouches using a straw. With everything floating, drinking from a glass could get very messy!

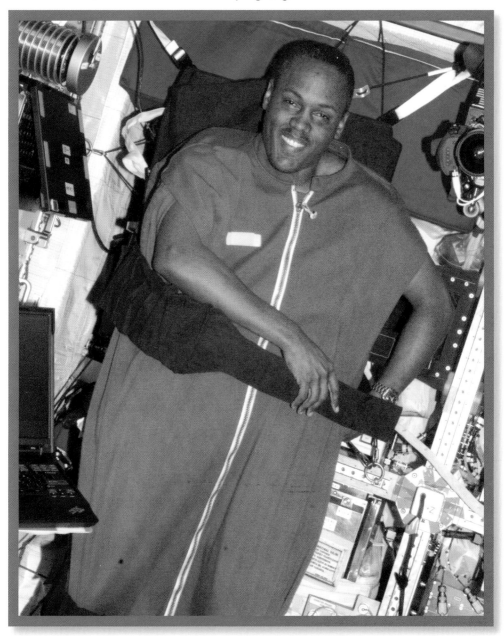

Most astronauts sleep in sleeping bags. They are tied to the wall so they won't float away.

Sometimes astronauts need to go outside the space station to do repairs or attach new parts. When they go out into space, the astronauts wear a special spacesuit called an *extravehicular mobility unit*, or EMU.

The EMU controls and monitors the astronaut's body temperature and breathing. It has a headphone and microphone so the astronaut can communicate with astronauts in the space station.

 A special jetpack is attached to the EMU. This jetpack can propel an astronaut freely through space. If an astronaut became separated from the space station, the jetpack could help him or her to get back safely.

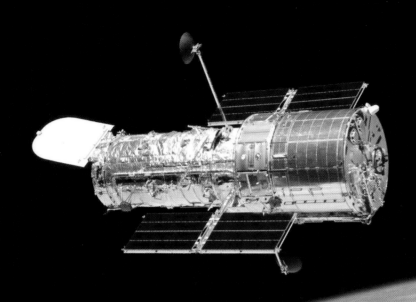

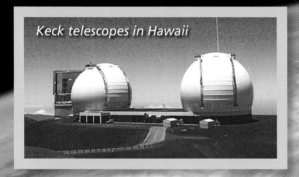

*Keck telescopes in Hawaii*

*Hubble telescope above Earth*

The International Space Station is just one of the many tools used to learn more about space. Another important tool for space exploration is the telescope. Some huge telescopes, like the Keck telescopes in Hawaii, are here on Earth. Others, like the Hubble telescope, are up in space. These telescopes allow special scientists, called **astronomers**, to see other **galaxies** of stars that are deep in space.

*Spiral galaxy M106*

**Astronomers** now believe there are billions of **galaxies** in the universe. Do you think there may be other forms of life in a far-off galaxy?

*Sombrero galaxy. A bright ring of dust encircles the galaxy.*

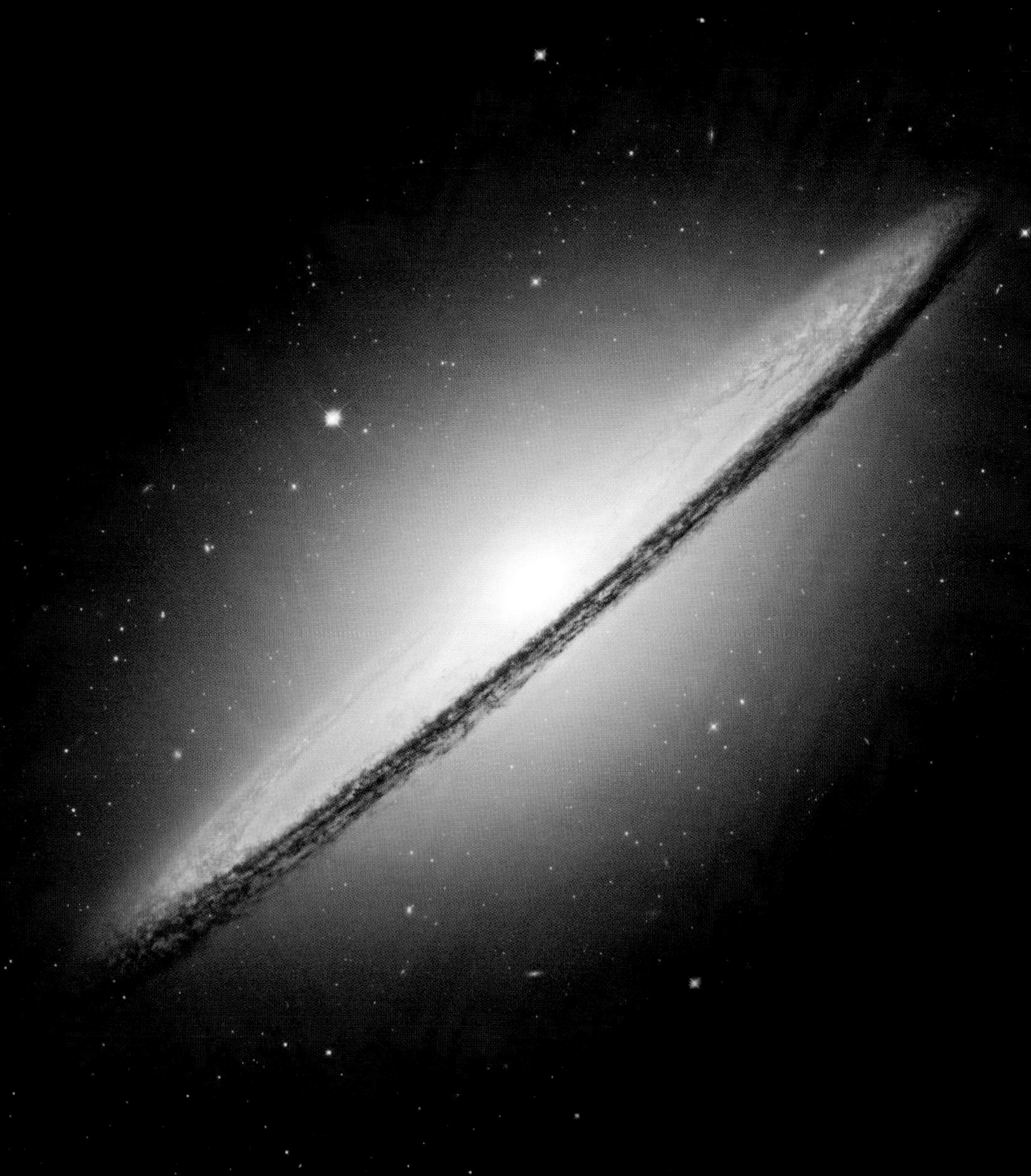

The galaxy pictured here is called the Sombrero galaxy. At the center of this galaxy is a huge black hole, believed to be a billion times more massive than our sun. The gravity in the black hole pulls things toward it.

*Illustration of a black hole (artist's concept)*

The gravity in a black hole is very strong. It sucks in everything around it. Even light cannot escape from a black hole.

*Small Magellanic Cloud, a small galaxy about 200,000 light-years from Earth*

There is so much more **exciting** information to learn about space. Maybe someday you will be a scientist or an astronomer. Perhaps you will make new discoveries and explore distant galaxies.

*Illustration of walk on Mars (artist's concept)*

What if you were an astronaut? Think of all the **exciting** things you could do. Maybe you could fly to Mars and help build a space station there!

# Glossary

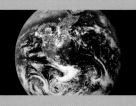

**atmosphere**
the mass of air that surrounds Earth

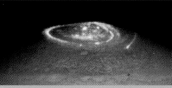

**aurora**
electrical and colorful light in the sky near the north or south poles

**galaxy**
any one of the large groups of stars that make up the universe

**satellite**
a large object sent into space that moves around Earth

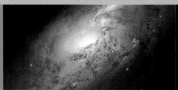

**solar system**
a sun and the planets that move around it

**telescope**
a device that makes distant objects appear larger

**universe**
all of space including stars, planets, and galaxies

# Questions to Ask after Reading

Add to the benefits of reading this book by discussing answers to these questions. Also consider discussing a few of your own questions.

**1** Why is the sun so important to us?

**2** Why is Venus called the "evening star?" Can you find the page that supports your answer?

**3** What kind of training do you think you would need to be an astronaut?

**4** Would you like to live in space? Why or why not?

**5** How would living in space be different than living on Earth?

**6** Can you tell me something that you would like to know about space or astronauts? How might you find information about this?

# Websites about Space

For more information about space,
you might want to check out these websites:

**www.spaceplace.nasa.gov**
**www.nasa.gov/audience/forstudents**
**www.windows2universe.org**

*Please note that these website addresses may no longer
be available. We recommend that children are always
supervised by a teacher or parent while on the Internet.*

If you liked ***About Space*** here are some other We Both Read® books you are sure to enjoy!

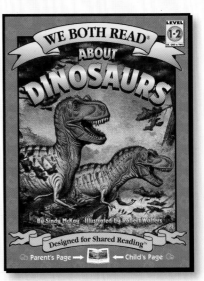

You can see all the We Both Read books that are available
at **WeBothRead.com**.